Let's Keep in Touch

Follow Us   Online

Visit US at

www.EffortlessMath.com

 https://www.facebook.com/Effortlessmath

Call

 https://goo.gl/2B6qWW

1-469-230-3605

Online Math Lessons

It's easy! Here's how it works.

1- Request a FREE introductory session.

2- Meet a Math tutor online via Skype.

3- Start Learning Math in Minutes.

Send Email to: info@EffortlessMath.com

Or Call: **+1-469-230-3605**

www.EffortlessMath.com

... So Much More Online!

- **FREE Math lessons**

- **More Math learning books!**

- **Online Math Tutors**

Looking for an Online Math Tutor?

Call us at: 001-469-230-3605

Send email to: Info@EffortlessMath.com

4th Grade SBAC Math Workbook 2018

The Most Comprehensive Review for the Math Section of the SBAC TEST

By

Reza Nazari

& Ava Ross

Copyright © 2018

Reza Nazari & Ava Ross

All rights reserved. No part of this publication may be reproduced, stored in a retrieval system, or transmitted in any form or by any means, electronic, mechanical, photocopying, recording, scanning, or otherwise, except as permitted under Section 107 or 108 of the 1976 United States Copyright Ac, without permission of the author.

All inquiries should be addressed to:

info@effortlessMath.com

www.EffortlessMath.com

ISBN-13: 978-1985316706

ISBN-10: 1985316706

Published by: Effortless Math Education

www.EffortlessMath.com

Description

Effortless Math SBAC Workbook provides students with the confidence and math skills they need to succeed on the SBAC Math, providing a solid foundation of basic Math topics with abundant exercises for each topic. It is designed to address the needs of SBAC test takers who must have a working knowledge of basic Math.

This comprehensive workbook with over 1,500 sample questions and 2 complete 4th Grade SBAC tests is all your student needs to fully prepare for the SBAC Math. It will help your student learns everything they need to ace the math section of the SBAC.

Effortless Math unique study program provides your student with an in-depth focus on the math portion of the exam, helping them master the math skills that students find the most troublesome.
This workbook contains most common sample questions that are most likely to appear in the mathematics section of the SBAC.

Inside the pages of this comprehensive workbook, students can learn basic math operations in a structured manner with a complete study program to help them understand essential math skills. It also has many exciting features, including:

- Dynamic design and easy-to-follow activities
- A fun, interactive and concrete learning process
- Targeted, skill-building practices
- Fun exercises that build confidence
- Math topics are grouped by category, so the students can focus on the topics they struggle on
- All solutions for the exercises are included, so you will always find the answers
- 2 Complete SBAC Math Practice Tests that reflect the format and question types on SBAC

Effortless Math SBAC Workbook is an incredibly useful tool for those who want to review all topics being covered on the SBAC test. It efficiently and effectively reinforces learning outcomes through engaging questions and repeated practice, helping students to quickly master basic Math skills.

About the Author

Reza Nazari is the author of more than 100 Math learning books including:
- **Math for Super Smart Students:** Fifth Graders and Older by Reza Nazari
- **Math and Critical Thinking Challenges:** For the Middle and High School Student
- **Effortless Math Education Workbooks**

Reza is also an experienced Math instructor and a test-prep expert who has been tutoring students since 2008. Reza is the founder of Effortless Math L.L.C, a tutoring company that has helped many students raise their standardized test scores--and attend the colleges of their dreams. Reza provides an individualized custom learning plan and the personalized attention that makes a difference in how students view math.

To ask questions about Math, you can contact Reza via email at:
reza@EffortlessMath.com

Find Reza's professional profile at:
https://goo.gl/zoC9rJ

Contents

Description 2

Place Values 8

Comparing and Ordering Numbers 9

Addition 10

Subtraction 11

Multiplication 12

Long Division by One Digit 13

Division with Remainders 14

Rounding and Estimating 15

Odd or Even 16

Decimal Place Value 17

Order and Comparing Decimals 18

Decimal Addition 19

Decimal Subtraction 20

Money Subtraction 21

Divisibility Rules 22

Fraction 23

Add Fractions with Like Denominators 24

Subtract Fractions with Like Denominators 25

Add and Subtract Fractions with Like Denominators 26

Compare Sums and Differences of Fractions with Like Denominators 27

Add 3 or More Fractions with Like Denominators 28

Simplifying Fractions 29

Add fractions with unlike denominators 30

Subtract fractions with unlike denominators 31

Add fractions with denominators of 10 and 100 32

Add and subtract fractions with denominators of 10, 100, and 1000 33

Fractions to Mixed Numbers 34

Mixed Numbers to Fractions 35

Add and Subtract Mixed Numbers with Like Denominators 36

Order of Operations 37

Line Segments 38

Identify lines of symmetry 39

Count lines of symmetry 40

Parallel, Perpendicular and Intersecting Lines 41

Graph points on a coordinate plane 42

Identifying Angles: Acute, Right, Obtuse, and Straight Angles 43

Angles of 90, 180, 270, and 360 Degrees 44

Measure Angles with a Protractor 45

Estimate Angle Measurements 46

Polygon Names 47

Classify triangles 48

Parallel Sides in Quadrilaterals 49

Identify Parallelograms 50

Identify Trapezoids 51

Identify Rectangles 52

Identify Rhombuses 53

Classify Quadrilaterals 54

Identify Three-Dimensional Figures 55

Count Vertices, Edges, and Faces 56

Identify Faces of Three-Dimensional Figures 57

Telling Time 58

Tally and Pictographs 59

Bar Graphs 60

Line Graphs 61

Patterns: Numbers 62

Perimeter 63

Perimeter: Find the Missing Side Lengths 64

Perimeter and Area of Squares 65

Perimeter and Area of rectangles 66

Find the Area or Missing Side Length of a Rectangle 67

Area of Complex Figures (With All Right Angles) 68

Area and Perimeter: Word Problems 69

Calculate Radius, Diameter, and Circumference 70

Measurement - Time 71

Measurement - Metric System 72

Measurement – Length 73

Measurement - Volume 74

Measurement - Temperature 75

Statistics 76

Roman Numerals 77

Patterns 78

SBAC Math Practice Tests 79

SBAC Practice Test 1 80

SBAC Practice Test 1 103

SBAC Practice Tests Answer Keys 122

Answers of the worksheets 123

Place Values

Write numbers in expanded form.

1) Thirty-five   30 + 5

2) Sixty-seven   ___ + ___

3) Forty-two   ___ + ___

4) Eighty-nine   ___ + ___

5) Ninety-one   ___ + ___

Circle the correct choice.

6) The 2 in 72 is in the   ones place   tens place   hundreds place

7) The 6 in 65 is in the   ones place   tens place   hundreds place

8) The 2 in 342 is in the   ones place   tens place   hundreds place

9) The 5 in 450 is in the   ones place   tens place   hundreds place

10) The 3 in 321 is in the   ones place   tens place   hundreds place

Comparing and Ordering Numbers

Use less than, equal to or greater than.

1) 23 _____ 34

2) 89 _____ 98

3) 45 _____ 25

4) 34 _____ 32

5) 91 _____ 91

6) 57 _____ 55

7) 85 _____ 78

8) 56 _____ 43

9) 34 _____ 34

10) 92 _____ 98

11) 38 _____ 46

12) 67 _____ 58

13) 88 _____ 69

14) 23 _____ 34

Order each set numbers from least to greatest.

15) $-15, -19, 20, -4, 1$ ___, ___, ___, ___, ___, ___

16) $6, -5, 4, -3, 2$ ___, ___, ___, ___, ___, ___

17) $15, -42, 19, 0, -22$ ___, ___, ___, ___, ___, ___

18) $26, -91, 0, -13, 67, -55$ ___, ___, ___, ___, ___, ___

19) $-17, -71, 90, -25, -54, -39$ ___, ___, ___, ___, ___, ___

20) $98, 5, 46, 19, 77, 24$ ___, ___, ___, ___, ___, ___

Addition

Add.

1) 1,158 + 6,687

2) 5,188 + 1,298

3) 5,756 + 2,712

4) 3,239 + 2,562

5) 4,257 + 5,194

6) 6,215 + 2,189

Find the missing numbers.

7) 45 + ___ = 105

8) 500 + 1000 = ___

9) 3200 + ___ = 4300

10) 455 + ___ = 1755

11) ___ + 720 = 1250

12) ___ + 670 = 2230

13) David sells gems. He finds a diamond in Istanbul and buys it for R3,433. Then, he flies to Cairo and purchases a bigger diamond for the bargain price of R5,922. How much does David spend on the two diamonds?

Subtraction

Subtract.

1) 8,519 − 5,422 = _____

2) 6,222 − 4,331 = _____

3) 7,821 − 3,212 = _____

4) 8,756 − 6,712 = _____

5) 9,290 − 3,829 = _____

6) 5,117 − 4,216 = _____

Find the missing number.

7) 2223 − ___ = 1120

8) 574 − ___ = 245

9) 1124 − 578 = ___

10) 2300 − ___ = 1250

11) 3780 − 1890 = ___

12) 2880 − 2560 = ___

13) Jackson had R3,963 invested in the stock market until he lost R2,171 on those investments. How much money does he have in the stock market now?

Multiplication

Find the answers.

1) 45 × 13 = _____

2) 32 × 10 = _____

3) 120 × 9 = _____

4) 563 × 4 = _____

5) 365 × 5 = _____

6) 89 × 25 = _____

7) 478 × 34 = _____

8) 956 × 26 = _____

9) 391 × 78 = _____

10) The Haunted House Ride runs 5 times a day. It has 6 cars, each of which can hold 4 people. How many people can ride the Haunted House Ride in one day?

11) Each train car has 3 rows of seats. There are 4 seats in each row. How many seats are there in 5 train cars?

Long Division by One Digit

Find the quotient.

1) $6\overline{)792}$

2) $5\overline{)350}$

3) $6\overline{)174}$

4) $8\overline{)104}$

5) $3\overline{)102}$

6) $9\overline{)189}$

7) $5\overline{)115}$

8) $2\overline{)120}$

9) $7\overline{)112}$

10) $4\overline{)148}$

11) $9\overline{)126}$

12) $6\overline{)240}$

13) $4\overline{)576}$

14) $4\overline{)512}$

15) $9\overline{)1278}$

16) $8\overline{)2768}$

17) $6\overline{)1224}$

18) $4\overline{)3412}$

Division with Remainders

Find the quotient with remainder.

1) $5 \overline{)592}$

2) $3 \overline{)295}$

3) $6 \overline{)553}$

4) $5 \overline{)214}$

5) $3 \overline{)440}$

6) $7 \overline{)673}$

7) $4 \overline{)213}$

8) $2 \overline{)820}$

9) $5 \overline{)496}$

10) $6 \overline{)791}$

11) $4 \overline{)647}$

12) $7 \overline{)780}$

13) $4 \overline{)5910}$

14) $8 \overline{)3515}$

15) $7 \overline{)2355}$

16) $9 \overline{)1232}$

17) $8 \overline{)6029}$

18) $4 \overline{)6743}$

Rounding and Estimating

Round each number to the underlined place value.

1) 9̲72

2) 2,9̲95

3) 36̲4

4) 8̲1

5) 5̲5

6) 33̲4

7) 1,2̲03

8) 9.5̲7

9) 7.4̲84

Estimate the sum by rounding each added to the nearest ten.

1) 55 + 9

2) 13 + 74

3) 83 + 7

4) 32 + 37

5) 13 + 74

6) 34 + 11

7) 39 + 77

8) 25 + 4

9) 61 + 73

10) 64 + 59

11) 14 + 68

12) 82 + 12

13) 43 + 66

14) 45 + 65

15) 553 + 232

Odd or Even

Identify whether each number is even or odd.

1) 12 _____

2) 7 _____

3) 33 _____

4) 18 _____

5) 99 _____

6) 55 _____

7) 34 _____

8) 87 _____

9) 94 _____

10) 14 _____

11) 22 _____

12) 79 _____

Circle the <u>even</u> number in each group.

13) 22, 11, 57, 13, 19, 47

14) 15, 17, 27, 23, 33, 26

15) 19, 35, 24, 56, 65, 49

16) 67, 58, 89, 63, 27, 63

Circle the <u>odd</u> number in each group.

17) 12, 14, 22, 64, 53, 98

18) 16, 26, 28, 44, 62, 73

19) 46, 82, 63, 98, 64, 56

20) 27, 92, 58, 36, 38, 72

Decimal Place Value

What place is the selected digit?

1) 1,12<u>2</u>.25

2) 2,321.3<u>2</u>

3) 4,258.91

4) 6,3<u>7</u>2.67

5) 7,131.<u>9</u>8

6) <u>5</u>,442.73

7) 1,841.8<u>9</u>

8) 5,995.<u>7</u>6

9) 8,<u>9</u>82.55

10) 1,24<u>9</u>.21

What is the value of the selected digit?

11) 3,122.3<u>1</u>

12) 1,3<u>1</u>8.66

13) 6,352.<u>25</u>

14) 3,<u>7</u>39.16

15) 4,9<u>3</u>6.78

16) 7,62<u>5</u>.86

17) 9,313.4<u>5</u>

18) <u>2</u>,168.82

19) 8,<u>4</u>51.76

20) 2,153.<u>23</u>

Order and Comparing Decimals

Use > = <.

1) 0.23 __ 0.34
2) 0.31 __ 0.37
3) 0.55 __ 0.47
4) 0.57 __ 0.59
5) 0.56 __ 0.67
6) 0.7 __ 0.67
7) 0.96 __ 8.55
8) 0.59 __ 0.88

9) 0.5 __ 0.25
10) 0.6 __ 0.3
11) 0.75 __ 0.6
12) 0.8 __ 0.80
13) 0.59 __ 0.6
14) 0.57 __ 0.75
15) 0.9 __ 0.11
16) 0.40 __ 0.4

Order each set of integers from least to greatest.

17) 0.4, 0.54, 0.23, 0.87, 0.36   ___, ___, ___, ___, ___, ___

18) 1.2, 2.4, 1.97, 3.65, 1.80   ___, ___, ___, ___, ___, ___

19) 2.3, 1.2, 1.9, 0.67, 0.34   ___, ___, ___, ___, ___, ___

20) 1.7, 1.2, 3.2, 4.2, 1.34, 3.55   ___, ___, ___, ___, ___, ___

Decimal Addition

Add.

1) 8.12 + 5.24 =

2) 1.5 + 1. 3 =

3) 7.2 + 1.34 =

4) 3. 4 + 1.75 =

5) 2.55 + 5.25 =

6) 5.78 + 4.30 =

7) 12.45 + 14.25 =

8) 13.67 + 11.31 =

9) 16.25 + 12.34 =

10) 10.25 + 12. 55 =

11) 21.25 + 20.90 =

12) 16.25 + 12.88 =

13) 18.44 + 12.65 =

14) 32.2 + 20.45 =

15) 15.76 + 15.98 =

16) 25.5 + 23.9 =

17) 30.95 + 21.40 =

18) 23.6 + 21.6 =

Decimal Subtraction

Subtract.

1) 6. 2 - 3.54 =

2) 5. 77 - 4.32 =

3) 8.66 - 6.55 =

4) 7.34 - 3.22 =

5) 4. 5 - 2.1 =

6) 3.78 - 2.55 =

7) 5.98 - 4.44 =

8) 4. 23 – 3.9 =

9) 16.5 – 13.12 =

10) 18.67 – 11.35 =

11) 12.98 – 10.45 =

12) 14.2 – 12. 4 =

13) 20.14 – 18.2 =

14) 25.6 – 24.2 =

15) 21.88 – 20.12 =

16) 27.55 – 23.4 =

17) 31.34 – 27.21 =

18) 23.34 – 21.5 =

Money Subtraction

Subtract.

1) $825 − $166 $651 − $110 $754 − $565

2) $539 − $137 $498 − $359 $992 − $549

3) $436 − $219 $512 − $128 $632 − $444

4) Linda had $12.00. She bought some game tickets for $7.14. How much did she have left?

Divisibility Rules

Use the divisibility rules to underline the factors of each number.

8	<u>2</u> 3 <u>4</u> 5 6 7 <u>8</u> 9 10
1) 16	2 3 4 5 6 7 8 9 10
2) 10	2 3 4 5 6 7 8 9 10
3) 15	2 3 4 5 6 7 8 9 10
4) 28	2 3 4 5 6 7 8 9 10
5) 36	2 3 4 5 6 7 8 9 10
6) 15	2 3 4 5 6 7 8 9 10
7) 27	2 3 4 5 6 7 8 9 10
8) 70	2 3 4 5 6 7 8 9 10
9) 57	2 3 4 5 6 7 8 9 10
10) 102	2 3 4 5 6 7 8 9 10
11) 144	2 3 4 5 6 7 8 9 10
12) 75	2 3 4 5 6 7 8 9 10

Fraction

What fraction of the squares is shaded?

1)

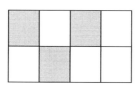

2)

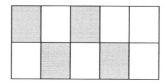

3)

Which fraction has the least value?

4) $\dfrac{1}{3}$ $\dfrac{2}{7}$ $\dfrac{8}{21}$ $\dfrac{4}{42}$

5) $\dfrac{1}{2}$ $\dfrac{3}{8}$ $\dfrac{3}{4}$ $\dfrac{9}{16}$

Add Fractions with Like Denominators

Add fractions.

1) $\dfrac{2}{3} + \dfrac{1}{3}$

2) $\dfrac{3}{5} + \dfrac{2}{5}$

3) $\dfrac{5}{8} + \dfrac{4}{8}$

4) $\dfrac{3}{4} + \dfrac{3}{4}$

5) $\dfrac{4}{10} + \dfrac{3}{10}$

6) $\dfrac{3}{7} + \dfrac{2}{7}$

7) $\dfrac{4}{5} + \dfrac{4}{5}$

8) $\dfrac{5}{14} + \dfrac{7}{14}$

9) $\dfrac{5}{18} + \dfrac{11}{18}$

10) $\dfrac{3}{12} + \dfrac{5}{12}$

11) $\dfrac{5}{13} + \dfrac{5}{13}$

12) $\dfrac{8}{25} + \dfrac{12}{25}$

13) $\dfrac{9}{15} + \dfrac{6}{15}$

14) $\dfrac{4}{20} + \dfrac{5}{20}$

15) $\dfrac{9}{17} + \dfrac{3}{17}$

16) $\dfrac{18}{32} + \dfrac{15}{32}$

17) $\dfrac{12}{28} + \dfrac{10}{28}$

18) $\dfrac{4}{20} + \dfrac{8}{20}$

19) $\dfrac{24}{45} + \dfrac{11}{45}$

20) $\dfrac{8}{36} + \dfrac{18}{36}$

21) $\dfrac{19}{30} + \dfrac{12}{30}$

Subtract Fractions with Like Denominators

Subtract fractions.

1) $\dfrac{4}{5} - \dfrac{2}{5}$

2) $\dfrac{2}{3} - \dfrac{1}{3}$

3) $\dfrac{7}{9} - \dfrac{4}{9}$

4) $\dfrac{5}{6} - \dfrac{3}{6}$

5) $\dfrac{4}{10} - \dfrac{3}{10}$

6) $\dfrac{5}{7} - \dfrac{3}{7}$

7) $\dfrac{7}{8} - \dfrac{5}{8}$

8) $\dfrac{11}{13} - \dfrac{9}{13}$

9) $\dfrac{8}{10} - \dfrac{5}{10}$

10) $\dfrac{8}{12} - \dfrac{7}{12}$

11) $\dfrac{18}{21} - \dfrac{12}{21}$

12) $\dfrac{15}{19} - \dfrac{9}{19}$

13) $\dfrac{9}{25} - \dfrac{6}{25}$

14) $\dfrac{25}{32} - \dfrac{17}{32}$

15) $\dfrac{22}{27} - \dfrac{9}{27}$

16) $\dfrac{27}{30} - \dfrac{15}{30}$

17) $\dfrac{31}{33} - \dfrac{26}{33}$

18) $\dfrac{18}{28} - \dfrac{8}{28}$

19) $\dfrac{35}{40} - \dfrac{15}{40}$

20) $\dfrac{29}{35} - \dfrac{19}{35}$

21) $\dfrac{21}{36} - \dfrac{11}{36}$

Add and Subtract Fractions with Like Denominators

Add fractions.

1) $\dfrac{1}{3} + \dfrac{2}{3}$

2) $\dfrac{3}{6} + \dfrac{2}{6}$

3) $\dfrac{5}{8} + \dfrac{2}{8}$

4) $\dfrac{3}{9} + \dfrac{5}{9}$

5) $\dfrac{4}{10} + \dfrac{1}{10}$

6) $\dfrac{3}{7} + \dfrac{2}{7}$

7) $\dfrac{3}{5} + \dfrac{2}{5}$

8) $\dfrac{1}{12} + \dfrac{1}{12}$

9) $\dfrac{16}{25} + \dfrac{5}{25}$

Subtract fractions.

10) $\dfrac{4}{5} - \dfrac{2}{5}$

11) $\dfrac{5}{7} - \dfrac{3}{7}$

12) $\dfrac{3}{4} - \dfrac{2}{4}$

13) $\dfrac{8}{9} - \dfrac{3}{9}$

14) $\dfrac{6}{14} - \dfrac{3}{14}$

15) $\dfrac{4}{15} - \dfrac{1}{15}$

16) $\dfrac{15}{16} - \dfrac{13}{16}$

17) $\dfrac{25}{50} - \dfrac{20}{50}$

18) $\dfrac{10}{21} - \dfrac{7}{21}$

Compare Sums and Differences of Fractions with Like Denominators

Evaluate and compare. Write < or >.

1) $\frac{1}{4} + \frac{2}{4} \underline{\quad} \frac{1}{4}$

2) $\frac{3}{5} + \frac{2}{5} \underline{\quad} \frac{4}{5}$

3) $\frac{5}{7} - \frac{3}{7} \underline{\quad} \frac{6}{7}$

4) $\frac{9}{10} + \frac{7}{10} \underline{\quad} \frac{5}{10}$

5) $\frac{5}{9} - \frac{3}{9} \underline{\quad} \frac{7}{9}$

6) $\frac{10}{12} - \frac{5}{12} \underline{\quad} \frac{3}{12}$

7) $\frac{3}{8} + \frac{1}{8} \underline{\quad} \frac{1}{8}$

8) $\frac{10}{15} + \frac{4}{15} \underline{\quad} \frac{9}{15}$

9) $\frac{15}{18} - \frac{3}{18} \underline{\quad} \frac{17}{18}$

10) $\frac{17}{21} + \frac{4}{21} \underline{\quad} \frac{18}{21}$

11) $\frac{14}{16} - \frac{4}{16} \underline{\quad} \frac{12}{16}$

12) $\frac{27}{32} - \frac{11}{32} \underline{\quad} \frac{20}{32}$

13) $\frac{25}{30} + \frac{5}{30} \underline{\quad} \frac{15}{30}$

14) $\frac{25}{27} - \frac{3}{27} \underline{\quad} \frac{9}{27}$

15) $\frac{42}{45} - \frac{15}{45} \underline{\quad} \frac{30}{45}$

16) $\frac{32}{36} + \frac{15}{36} \underline{\quad} \frac{18}{36}$

Add 3 or More Fractions with Like Denominators

Add fractions.

1) $\frac{4}{7} + \frac{2}{7} + \frac{1}{7}$

2) $\frac{1}{5} + \frac{3}{5} + \frac{1}{5}$

3) $\frac{3}{9} + \frac{3}{9} + \frac{1}{9}$

4) $\frac{1}{4} + \frac{1}{4} + \frac{1}{4}$

5) $\frac{7}{15} + \frac{3}{15} + \frac{4}{15}$

6) $\frac{3}{12} + \frac{2}{12} + \frac{3}{12}$

7) $\frac{4}{10} + \frac{2}{10} + \frac{1}{10}$

8) $\frac{5}{18} + \frac{5}{18} + \frac{3}{18}$

9) $\frac{5}{21} + \frac{11}{21} + \frac{3}{21}$

10) $\frac{2}{16} + \frac{5}{16} + \frac{8}{16}$

11) $\frac{4}{25} + \frac{4}{25} + \frac{4}{25}$

12) $\frac{12}{30} + \frac{7}{30} + \frac{5}{30}$

13) $\frac{9}{27} + \frac{6}{27} + \frac{6}{27}$

14) $\frac{3}{42} + \frac{5}{42} + \frac{6}{42}$

Simplifying Fractions

Simplify the fractions.

1) $\dfrac{22}{36}$

2) $\dfrac{8}{10}$

3) $\dfrac{12}{18}$

4) $\dfrac{6}{8}$

5) $\dfrac{13}{39}$

6) $\dfrac{5}{20}$

7) $\dfrac{16}{36}$

8) $\dfrac{18}{36}$

9) $\dfrac{50}{20}$

10) $\dfrac{6}{54}$

11) $\dfrac{45}{81}$

12) $\dfrac{21}{28}$

13) $\dfrac{35}{56}$

14) $\dfrac{52}{64}$

15) $\dfrac{13}{65}$

16) $\dfrac{44}{77}$

17) $\dfrac{21}{42}$

18) $\dfrac{15}{36}$

19) $\dfrac{9}{24}$

20) $\dfrac{20}{80}$

21) $\dfrac{25}{45}$

Add fractions with unlike denominators

Add fractions.

1) $\dfrac{2}{3} + \dfrac{4}{2}$

2) $\dfrac{3}{5} + \dfrac{2}{6}$

3) $\dfrac{5}{6} + \dfrac{4}{8}$

4) $\dfrac{7}{4} + \dfrac{5}{9}$

5) $\dfrac{4}{10} + \dfrac{1}{5}$

6) $\dfrac{3}{7} + \dfrac{2}{4}$

7) $\dfrac{3}{4} + \dfrac{2}{5}$

8) $\dfrac{2}{3} + \dfrac{1}{5}$

9) $\dfrac{16}{25} + \dfrac{34}{50}$

10) $\dfrac{2}{3} + \dfrac{1}{2}$

11) $\dfrac{3}{12} + \dfrac{2}{4}$

12) $\dfrac{5}{3} + \dfrac{1}{15}$

Subtract fractions with unlike denominators

Subtract fractions.

1) $\dfrac{4}{5} - \dfrac{2}{6}$

2) $\dfrac{2}{7} - \dfrac{3}{5}$

3) $\dfrac{1}{2} - \dfrac{2}{4}$

4) $\dfrac{8}{9} - \dfrac{3}{5}$

5) $\dfrac{3}{7} - \dfrac{3}{14}$

6) $\dfrac{4}{15} - \dfrac{1}{10}$

7) $\dfrac{12}{16} - \dfrac{13}{18}$

8) $\dfrac{25}{40} - \dfrac{20}{50}$

9) $\dfrac{1}{2} - \dfrac{1}{9}$

10) $\dfrac{13}{25} - \dfrac{1}{5}$

11) $\dfrac{1}{3} - \dfrac{1}{27}$

12) $\dfrac{22}{42} - \dfrac{2}{7}$

13) $\dfrac{27}{36} - \dfrac{5}{12}$

14) $\dfrac{17}{20} - \dfrac{2}{4}$

Add fractions with denominators of 10 and 100

Add fractions.

1) $\dfrac{5}{10} + \dfrac{20}{100}$

2) $\dfrac{2}{10} + \dfrac{35}{100}$

3) $\dfrac{25}{100} + \dfrac{6}{10}$

4) $\dfrac{73}{100} + \dfrac{1}{10}$

5) $\dfrac{68}{100} + \dfrac{2}{10}$

6) $\dfrac{4}{10} + \dfrac{40}{100}$

7) $\dfrac{80}{100} + \dfrac{1}{10}$

8) $\dfrac{50}{100} + \dfrac{3}{10}$

9) $\dfrac{59}{100} + \dfrac{3}{10}$

10) $\dfrac{7}{10} + \dfrac{12}{100}$

11) $\dfrac{9}{10} + \dfrac{10}{100}$

12) $\dfrac{40}{100} + \dfrac{3}{10}$

13) $\dfrac{36}{100} + \dfrac{4}{10}$

14) $\dfrac{27}{100} + \dfrac{6}{10}$

15) $\dfrac{55}{100} + \dfrac{3}{10}$

16) $\dfrac{1}{10} + \dfrac{85}{100}$

17) $\dfrac{17}{100} + \dfrac{6}{10}$

18) $\dfrac{26}{100} + \dfrac{7}{10}$

Add and subtract fractions with denominators of 10, 100, and 1000

Evaluate fractions.

1) $\frac{8}{10} - \frac{30}{100}$

2) $\frac{6}{10} + \frac{27}{100}$

3) $\frac{25}{100} + \frac{450}{1000}$

4) $\frac{73}{100} - \frac{320}{1000}$

5) $\frac{25}{100} + \frac{670}{1000}$

6) $\frac{4}{10} + \frac{780}{1000}$

7) $\frac{80}{100} - \frac{560}{1000}$

8) $\frac{78}{100} - \frac{6}{10}$

9) $\frac{820}{1000} + \frac{5}{10}$

10) $\frac{67}{100} + \frac{240}{1000}$

11) $\frac{7}{10} - \frac{12}{100}$

12) $\frac{75}{100} - \frac{5}{10}$

13) $\frac{70}{100} - \frac{3}{10}$

14) $\frac{850}{1000} - \frac{5}{100}$

15) $\frac{300}{1000} + \frac{12}{100}$

16) $\frac{780}{1000} - \frac{6}{10}$

17) $\frac{80}{100} - \frac{6}{10}$

18) $\frac{50}{100} - \frac{210}{1000}$

Fractions to Mixed Numbers

Convert fractions to mixed numbers.

1) $\dfrac{9}{4}$

2) $\dfrac{37}{5}$

3) $\dfrac{21}{6}$

4) $\dfrac{41}{10}$

5) $\dfrac{11}{2}$

6) $\dfrac{56}{10}$

7) $\dfrac{20}{12}$

8) $\dfrac{9}{5}$

9) $\dfrac{19}{5}$

10) $\dfrac{27}{10}$

11) $\dfrac{10}{6}$

12) $\dfrac{17}{8}$

13) $\dfrac{7}{2}$

14) $\dfrac{39}{4}$

15) $\dfrac{72}{10}$

16) $\dfrac{13}{3}$

17) $\dfrac{45}{8}$

18) $\dfrac{27}{5}$

Mixed Numbers to Fractions

Convert to fraction.

1) $1\frac{2}{6}$

2) $2\frac{2}{3}$

3) $5\frac{1}{3}$

4) $6\frac{4}{5}$

5) $2\frac{3}{4}$

6) $2\frac{5}{7}$

7) $3\frac{5}{9}$

8) $2\frac{9}{10}$

9) $7\frac{5}{6}$

10) $6\frac{11}{12}$

11) $8\frac{9}{20}$

12) $8\frac{2}{5}$

13) $5\frac{4}{5}$

14) $9\frac{1}{6}$

15) $3\frac{3}{4}$

16) $10\frac{2}{3}$

17) $12\frac{3}{4}$

18) $14\frac{6}{7}$

Add and Subtract Mixed Numbers with Like Denominators

Add mixed numbers.

1) $5\frac{8}{20} + 8\frac{10}{20}$

2) $4\frac{5}{10} + 6\frac{8}{10}$

3) $6\frac{2}{4} + 7\frac{1}{4}$

4) $5\frac{2}{6} + 8\frac{3}{6}$

5) $5\frac{1}{3} - 1\frac{2}{3}$

6) $7\frac{6}{20} - 1\frac{10}{20}$

7) $7\frac{5}{9} - 2\frac{7}{9}$

8) $4\frac{8}{10} - 2\frac{9}{10}$

9) $5\frac{23}{25} - 1\frac{24}{25}$

10) $6\frac{4}{6} + 4\frac{3}{6}$

11) $3\frac{3}{8} + 2\frac{1}{8}$

12) $6\frac{4}{10} + 2\frac{2}{10}$

13) $1\frac{2}{12} + 7\frac{4}{12}$

14) $2\frac{2}{9} + 3\frac{4}{9}$

Order of Operations

Evaluate each expression.

1) $(2 \times 2) + 5$

2) $24 - (3 \times 3)$

3) $(6 \times 4) + 8$

4) $25 - (4 \times 2)$

5) $(6 \times 5) + 3$

6) $64 - (2 \times 4)$

7) $25 + (1 \times 8)$

8) $(6 \times 7) + 7$

9) $48 \div (4 + 4)$

10) $(7 + 11) \div (-2)$

11) $9 + (2 \times 5) + 10$

12) $(5 + 8) \times \frac{3}{5} + 2$

13) $2 \times 7 - \frac{10}{9 - 4}$

14) $(12 + 2 - 5) \times 7 - 1$

15) $\frac{7}{5 - 1} \times (2 + 6) \times 2$

16) $20 \div (4 - (10 - 8))$

17) $\frac{50}{4(5 - 4) - 3}$

18) $2 + (8 \times 2)$

Line Segments

Write each as a line, ray or line segment.

1)

2)

3)

4)

5)

6)

7)

8)

Identify lines of symmetry

Tell whether the line on each shape is a line of symmetry.

1)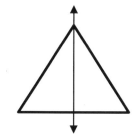

2)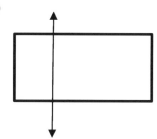

3)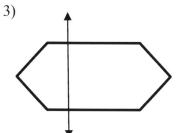

4)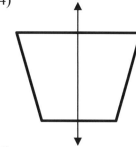

5)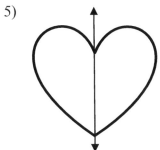

6)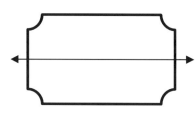

7)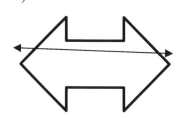

8)

Count lines of symmetry

Draw lines of symmetry on each shape. Count and write the lines of symmetry you see.

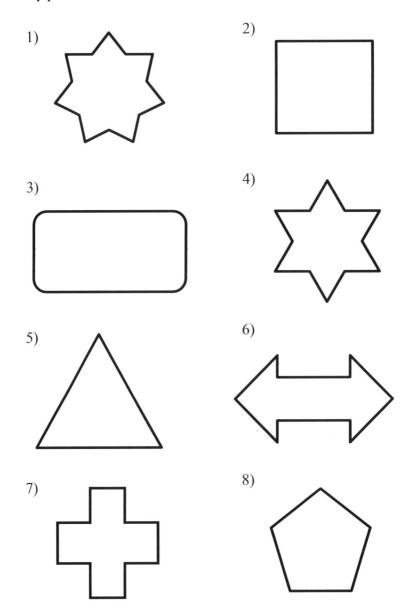

Parallel, Perpendicular and Intersecting Lines

State wheither the given pair of lines are parallel, perpendicular, or intersecting.

1)

2)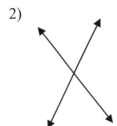

3)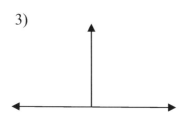

4)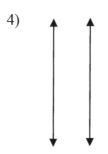

5)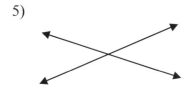

6)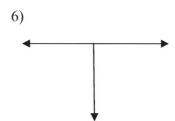

7)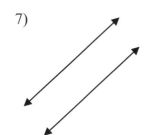

8)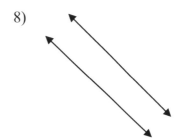

Graph points on a coordinate plane

Plot each point on the coordinate grid.

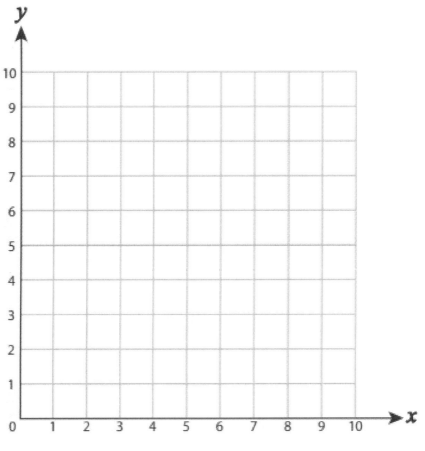

1) A (3, 6)   3) C (3, 7)   5) F (5, 2)

2) B (1, 3)   4) E (8, 6)   6) G (9, 3)

Identifying Angles: Acute, Right, Obtuse, and Straight Angles

Write the name of the angles.

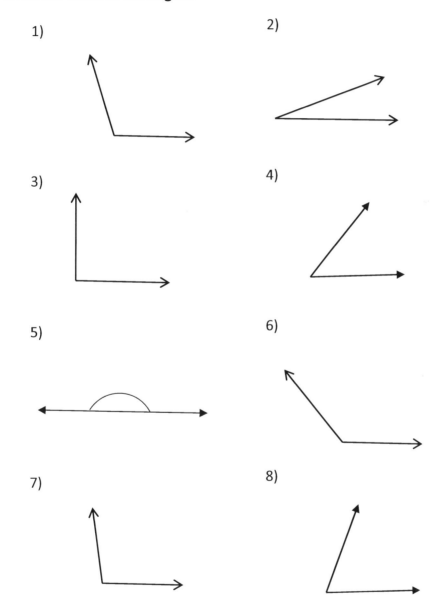

Angles of 90, 180, 270, and 360 Degrees

Measurement of each angle in degrees.

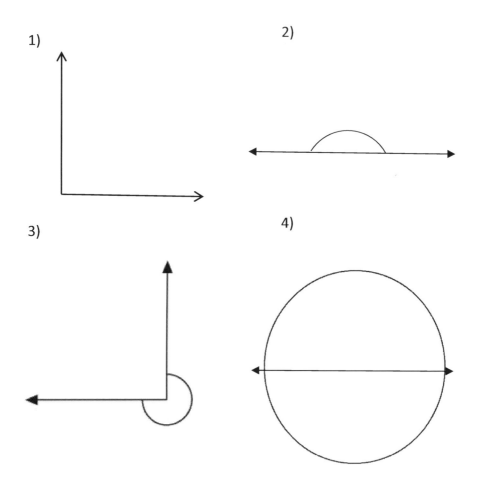

1)

2)

3)

4)

Measure Angles with a Protractor

Use a protractor to measure the angles below.

1)

2)

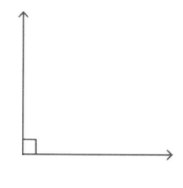

3)

4)

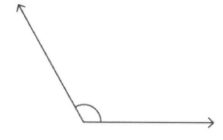

Use a protractor to draw angles for each measurement given.

5) 45°   6) 90°   7) 145°

Estimate Angle Measurements

Estimate the approximate measurement of each angle in degrees.

5)

6)

7)

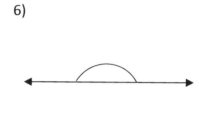

8)

9)

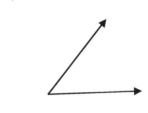

10)

11)

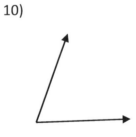

12)

Polygon Names

Write name of polygons.

1)

2)

3)
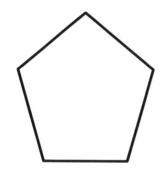

4)
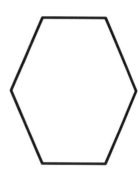

5)
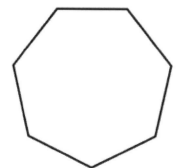

6)
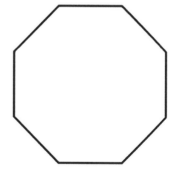

Classify triangles

Classify the triangles by their sides and angles.

1)
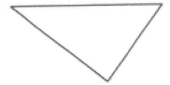

2)

3)
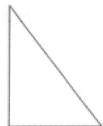

4)
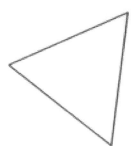

5)

6)
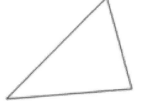

Parallel Sides in Quadrilaterals

Write name of quadrilaterals.

1)

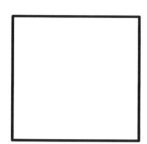

2)

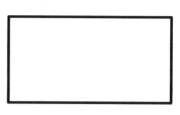

3)

4)

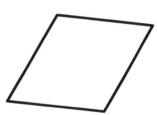

5)

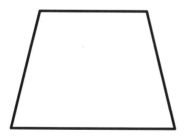

6)

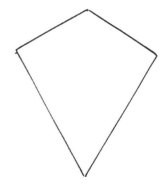

Identify Parallelograms

Write the name of parallelograms.

1)

2)

3)

4)

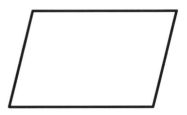

Identify Trapezoids

Which shapes is trapezoid?

1)

2)

3)

4)

5)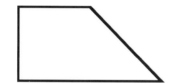

6)

7)

8)

Identify Rectangles

1) A rectangle has _____ sides and _____ angles.

2) Draw a rectangle that is 6 centimeters long and 3 centimeters wide. What is the perimeter?

3) Draw a rectangle 5 cm long and 2 cm wide..

4) Draw a rectangle whose length is 4 cm and whose width is 2 cm. What is the perimeter of the rectangle?

5) What is the perimeter of the rectangle?   8

   4

Identify Rhombuses

Which shape is a Rhombus?

1)
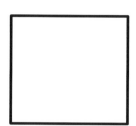

2)

3)
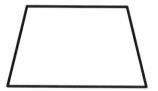

4)
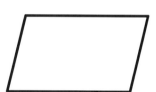

5)
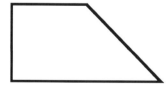

6)
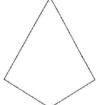

Classify Quadrilaterals

Identify the type for each quadrilateral.

7)

8)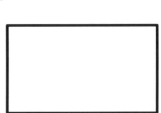

9)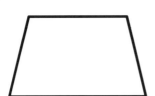

10)

11)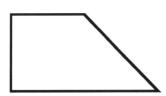

12)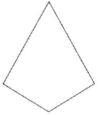

Identify Three-Dimensional Figures

Write the name of each shape.

1)

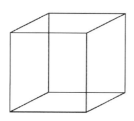

2)

3)

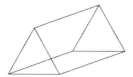

4)

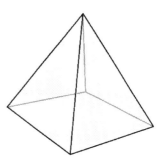

5)

6)

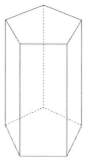

7)

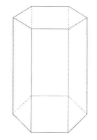

Count Vertices, Edges, and Faces

	Number of edges	Number of faces	Number of vertices
1)	_____	_____	_____
2)	_____	_____	_____
3)	_____	_____	_____
4)	_____	_____	_____
5)	_____	_____	_____
6)	_____	_____	_____

Identify Faces of Three-Dimensional Figures

Write the number of faces.

1)

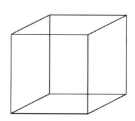

2)

3)

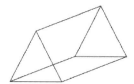

4)

5)

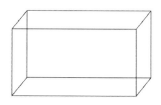

6)

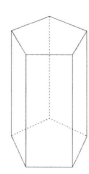

7)

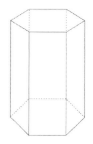

8)

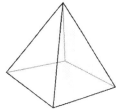

Telling Time

1) What time is shown by this clock?

   _____

2) It is night. What time is shown on this clock?

   _____

How much time has passed?

3) From 1:15 AM to 4:35 AM: _____ hours and _____ minutes.

4) From 2:35 AM to 5:10 AM: _____ hours and _____ minutes.

5) It's 8:30 P.M. What time was 5 hours ago?

   _____ O'clock

Tally and Pictographs

Using the key, draw the pictograph to show the information.

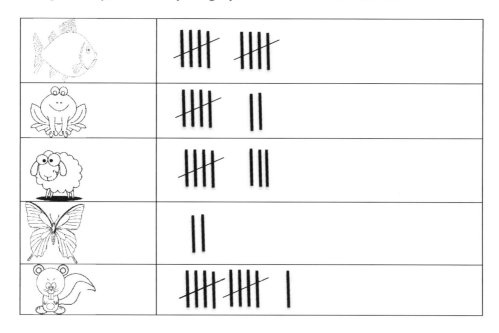

Key: ☺ = 2 animals

Bar Graphs

Graph the given information as a bar graph.

Day	Hot dogs sold
Monday	90
Tuesday	70
Wednesday	30
Thursday	20
Friday	60

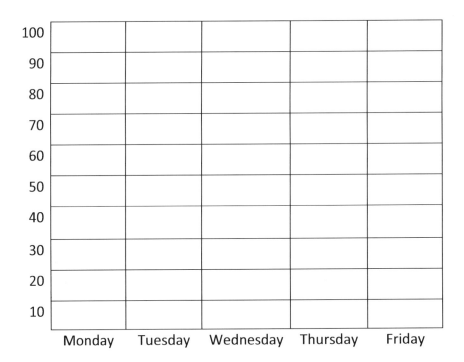

Line Graphs

David work as a salesman in a store. He records the number of shoes sold in five days on a line graph. Use the graph to answer the questions.

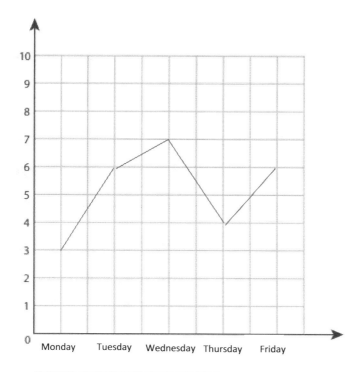

1) How many cars were sold on Monday?

2) Which day had the minimum sales of shoes?

3) Which day had the maximum number of shoes sold?

4) How many shoes were sold in 5 days?

Patterns: Numbers

Write the numbers that come next.

1) 3, 6, 9 , 12,   _____, _____, _____, _____

2) 2, 4, 6, 8,   _____, _____, _____, _____

3) 5, 10, 15, 20,   _____, _____, _____, _____

4) 15, 25, 35, 45,   _____, _____, _____, _____

5) 11, 22, 33, 44,   _____, _____, _____, _____

6) 10, 18, 26, 34, 8ta   _____, _____, _____, _____

7) 61, 55, 49, 43, 6 ta   _____, _____, _____, _____

8) 45, 56, 67, 78,   _____, _____, _____, _____

Perimeter

What is the perimeter of the shapes?

1)

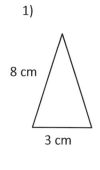

2)

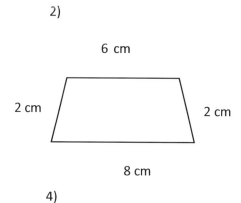

3)

4)

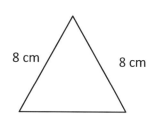

5)

Perimeter: Find the Missing Side Lengths

Find the side length of each square.

1) perimeter = 44

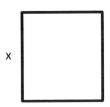

2) perimeter = 28

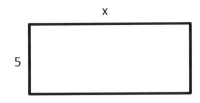

3) perimeter = 30

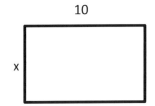

4) perimeter = 16

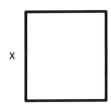

5) perimeter = 60

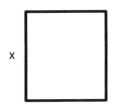

6) perimeter = 22

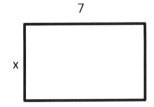

7) perimeter = 30

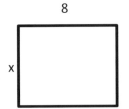

8) perimeter = 36

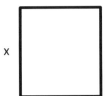

Perimeter and Area of Squares

Find perimeter and area of squares.

1) A: _____ , P: _____

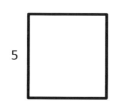
5

2) A: _____ , P: _____

3

3) A: _____ , P: _____

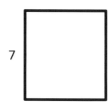
7

4) A: _____ , P: _____

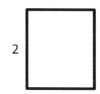
2

5) A: _____ , P: _____

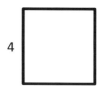
4

6) A: _____ , P: _____

10

7) A: _____ , P: _____

8

8) A: _____ , P: _____

12

Perimeter and Area of rectangles

Find perimeter and area of rectangles.

1) A: _____, P: _____

2) A: _____, P: _____

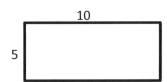

3) A: _____, P: _____

4) A: _____, P: _____

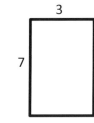

5) A: _____, P: _____

6) A: _____, P: _____

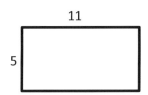

7) A: _____, P: _____

8) A: _____, P: _____

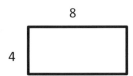

Find the Area or Missing Side Length of a Rectangle

Find area or missing side length of rectangles.

1) Area = ?

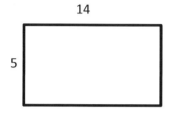

2) Area = 48, x = ?

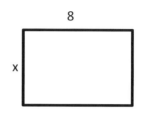

3) Area = 40, x = ?

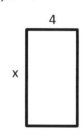

4) Area = ?

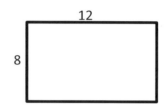

5) Area = ?

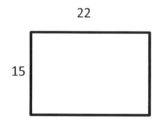

6) Area = 600, X = ?

7) Area = 384, x = ?

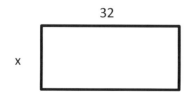

8) Area = 525, x = ?

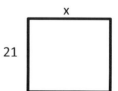

Area of Complex Figures (With All Right Angles)

Find area of shapes.

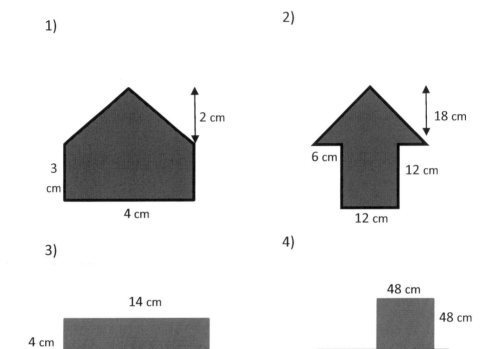

Area and Perimeter: Word Problems

1) The area of a rectangle is 72 square meters. The width is 8 meters. What is the length of the rectangle?

2) A square has an area of 36 square feet. What is the perimeter of the square?

3) Ava built a rectangular vegetable garden that is 6 feet long and has an area of 54 square feet. What is the perimeter of Ava's vegetable garden?

4) A square has a perimeter of 64 millimeters. What is the area of the square?

5) The perimeter of David's square backyard is 44 meters. What is the area of David's backyard?

6) The area of a rectangle is 40 square inches. The length is 8 inches. What is the perimeter of the rectangle?

Calculate Radius, Diameter, and Circumference

Find the diameter and circumference of circles.

1)
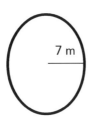
7 m

2)
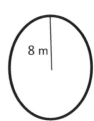
8 m

3)

12 m

4)

Find the radius.

5)
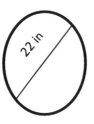
22 in

6)
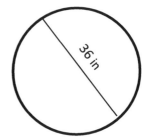
36 in

7) diameter = 46
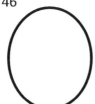

8) diameter = 24
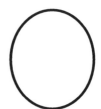

Measurement - Time

How much time has passed?

1) 1:15 AM to 4:35 AM: _____ hours and _____ minutes.

2) 2:35 AM to 5:10 AM: _____ hours and _____ minutes.

3) 6:00 AM. to 7:25 AM. = ___ hour(s) and ___ minutes.

4) 6:15 PM to 7:30 PM. = ___ hour(s) and ___ minutes

5) 5:15 a.m. to 5:45 a.m. = _____ minutes

6) 4:05 a.m. to 4:30 a.m. = _____ minutes

7) There are _____ second in 15 minutes.

8) There are _____ second in 11 minutes.

9) There are _____ second in 27 minutes.

10) There are _____ minutes in 10 hours.

11) There are _____ minutes in 20 hours.

12) There are _____ minutes in 12 hours.

Measurement - Metric System

Convert to the units.

1) 4 mm = _____ cm

2) 0.6 m = _____ mm

3) 2 m = _____ cm

4) 0.03 km = _____ m

5) 3000 mm = _____ km

6) 5 cm _____ m

7) 0.03 m = _____ cm

8) 1000 mm = _____ km

9) 600 mm = _____ m

10) 0.77 km = _____ mm

11) 0.08 km = _____ m

12) 0.30 m = _____ cm

13) 400 m = _____ km

14) 5000 cm = _____ km

15) 40 mm = _____ cm

16) 800 m = _____ km

Measurement – Length

1) Use a ruler to find the length of the line segment below to the nearest quarter inch.

_____

_____

_____

Convert the following measurements.

2) 2 feet: ____ inches

3) 5 feet ̂ ____ inches

4) 1 yard ̂ ____ feet

5) 3 yards ̂ ____ feet

6) 1 meter ̂ ____ centimeter

7) 3 kilometers ̂ ____ meters

8) 100 meters ̂ ____ centimeters

9) 8 yards ̂ ____ feet

Measurement - Volume

Find the volume of each of the rectangular prisms.

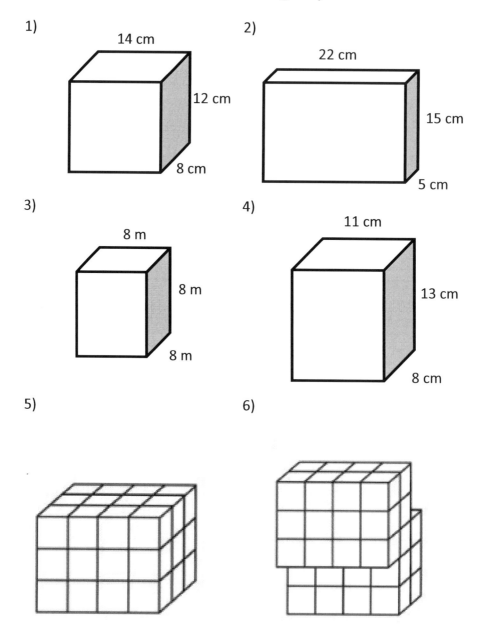

Measurement - Temperature

4) What temperature is shown on this Celsius thermometer? 20

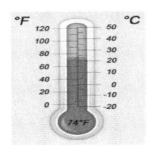

Convert Celsius into Fahrenheit.

1) 10°C = ___ °F

2) 35°C = ___ °F

3) 80°C = ___ °F

4) 15°C = ___ °F

5) 25°C = ___ °F

6) 50°C = ___ °F

7) 45°C = ___ °F

8) 90°C = ___ °F

9) 30°C = ___ °F

10) 20°C = ___ °F

11) 70°C = ___ °F

12) 80°C = ___ °F

Statistics

Write Mean, Median, Mode, and Range of the Given Data.

1) 7, 2, 5, 1, 1, 2

2) 2, 2, 2, 3, 6, 3, 7, 4

3) 9, 4, 3, 1, 7, 9, 4, 6, 4

4) 8, 4, 2, 4, 3, 2, 4, 5

5) 8, 5, 7, 5, 7, 9, 8

6) 5, 1, 4, 4, 9, 2, 9, 2, 5, 1

7) 4, 1, 5, 9, 7, 7, 5, 4, 3, 5

8) 7, 5, 4, 9, 6, 7, 7, 5, 2

9) 2, 5, 5, 6, 2, 4, 7, 6, 4, 9

10) 10, 5, 2, 5, 4, 5, 8, 10

11) 5, 1, 5, 2, 2

12) 2, 3, 5, 9, 6

Roman Numerals

Write in Romans numerals.

1. 2 _____ 2. 6 _____

3. 4 _____ 4. 9 _____

5. 10 _____ 6. 7 _____

7. 3 _____ 8. 1 _____

9. 5 _____ 10. 8 _____

11. Add 7 + 2 and write in Roman numerals. _____

12. Add 6 + 5 and write in Roman numerals. _____

13. Subtract 12 − 6 and write in Roman numerals. _____

14. Subtract 20 − 8 and write in Roman numerals. _____

15. Subtract 16 − 9 and write in Roman numerals. _____

16. Add 10 + 10 and write in Roman numerals. _____

Patterns

Write the next three numbers in each counting sequence.

1) -32, -23, -14, _____, _____, _____, _____

2) 543, 528, 513, _____, _____, _____, _____

3) _____, _____, 56, 64, _____, 80

4) 23, 34, _____, _____, 67, _____

5) 24, 31, _____, _____, _____

6) 52, 45, _____, _____, _____

7) 51, 44, 37, _____, _____, _____

8) 64, 51, 38, _____, _____, _____

9) What are the next three numbers in this counting sequence?

    1350, 1550, 1750, _____, _____, _____

10) What is the seventh number in this counting sequence?

    7, 16, 25, _____

SBAC Math Practice Tests

Smarter Balanced Assessment Consortium (SBAC) test assesses student mastery of the Common Core State Standards.

The SBAC is a computer adaptive test. It means that there is a set of test questions in a variety of question types that adjust to each student based on the student's answers to previous questions. This section includes a range of items types, such as selecting several correct responses for one item, typing out a response, fill---in short answers/tables, graphing, drag and drop, etc.

On computer adaptive tests, if the correct answer is chosen, the next question will be harder. If the answer given is incorrect, the next question will be easier. This also means that once an answer is selected on the computer it cannot be changed.

In this section, there are 2 complete SBAC Math Tests that reflect the format and question types on SBAC. On a real SBAC Math test, the number of questions varies and there are about 30 questions.

Let your student take these tests to see what score he or she will be able to receive on a real SBAC test.

SBAC Practice Test 1

Smarter Balanced Assessment Consortium

Grade 4

Mathematics

2018

- 30 questions
- There is no time limit for this practice test.
- Calculators are NOT permitted for this practice test

1. Erik made 12 pints of juice. He drinks 3 cups of juice each day. How many days will Erik take to drink all of the juice he made?

    A. 2 days

    B. 4 days

    C. 8 days

    D. 9 days

2. Jason has prepared $\frac{4}{10}$ of his assignment. Which decimal represent the part of the assignment Jason has prepared?

    A. 4.10

    B. 4.01

    C. 0.4

    D. 0.04

3. Emma described a number using these clues.
   - 3 digits of the number are 4, 7, and 9
   - The value of the digit 4 is (4 × 10)
   - The value of the digit 7 is (7 × 1000)
   - The value of the digit 9 is (9 × 10000)

   Which number could fit Emma's description?

   A. 9,724.04

   B. 90,734.40

   C. 97,040.04

   D. 98,740.70

4. There are 18 boxes and each box contains 26 pencils. How many pencils are in the boxed in total?

    A. 108

    B. 208

    C. 468

    D. Not here

5. Emily and Ava were working on a group project last week. They completed $\frac{7}{10}$ of their project on Tuesday and the rest on Wednesday. Ava completed $\frac{3}{10}$ of their project on Tuesday. What fraction of the group project did Emily completed on Tuesday?

    A. $\frac{2}{10}$

    B. $\frac{3}{10}$

    C. $\frac{4}{10}$

    D. $\frac{7}{10}$

6. Joe has 855 crayons. What is this number rounded to the nearest ten?

    Write your answer in the box below.

7. The length of the following rectangle is 4 inches and the width of the rectangle is 3 inches.

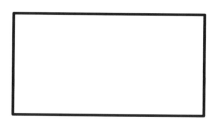

What is the area of the rectangle in square inches?

A. 12 square inches

B. 14 square inches

C. 18 square inches

D. square inches

8. Peter's pencil is $\frac{12}{100}$ of a meter long. What is the length, in meters, of Peter's pencil written as a decimal?

   A. 0.12

   B. 1.02

   C. 1.2

   D. 12.100

9. There are 7 days in a week. There are 28 days in the month of February. How many times as many days are there in February than are in one week?

   A. 4 times

   B. 7 times

   C. 21 times

   D. 35 times

10. A basketball team is buying new uniforms. Each uniform costs $24. The team wants to buy 14 uniforms.

    Which equation represents a way to find the total cost of the uniforms?

    A. (20 × 10) + (4 × 4) = 200 + 16

    B. (20 × 4) + (10 × 4) = 80 + 40

    C. (24 × 10) + (24 × 4) = 240 + 96

    D. (24 × 4) + (4 × 14) = 96 + 56

11. Use the table below to answer the question.

Favorite Sports

Sport	Number of Votes
football (FB)	10
basketball (BB)	5
soccer (SOC)	7
volleyball (VB)	2

The students in the fourth grade class voted for their favorite sport. Which bar graph shows results of the students vote?

A.
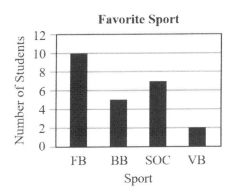

B.
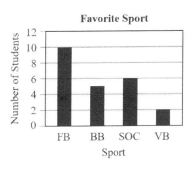

C.
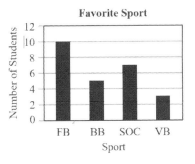

D.
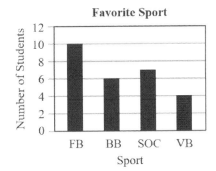

12. For a concert, there are children's tickets and adult tickets for sale. Of the total available tickets, $\frac{26}{100}$ have been sold as adult tickets and $\frac{4}{10}$ as children's tickets. The rest of the tickets have not been sold.

What fraction of the total number of tickets for the concert have been sold?

A. $\frac{30}{100}$

B. $\frac{66}{100}$

C. $\frac{30}{10}$

D. $\frac{66}{110}$

13. Mia has a group of shapes. Each shape in her group has at least one set of parallel sides. Each shape also has at least one set of perpendicular sides. Which group could be Mia's group of shapes?

A.
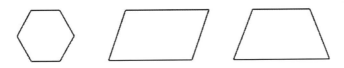

B.
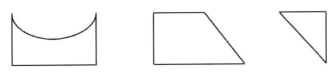

C.
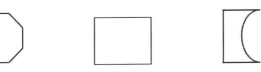

D.

14. There are 86 students from Riddle Elementary school at the library on Monday. The other 32 students in the school are practicing in the classroom. Which number sentence shows the total number of students in Riddle Elementary school?

   A. 86 + 32
   B. 86 − 32
   C. 86 × 32
   D. 86 ÷ 32

15. Use the picture below to answer the question.

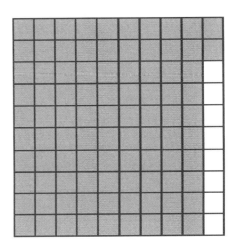

Which decimal number names the shaded part of this square?
   A. 0.08
   B. 0.20
   C. 0.92
   D. 0.98

16. Which number correctly completes the number sentence 80 × 34 = ?

- A. 272
- B. 560
- C. 1,920
- D. 2,720

17. Use the table below to answer the question.

City Populations

City	Population
Denton	28,097
Bomberg	28,207
Windham	29,700
Sanhill	27,980

Which list of city populations is in order from least to greatest?

A. 28,097;   28,207; 29,700; 27,980

B. 29,700;   28,207; 28,097; 27,980

C. 27,980;   28,097; 28,207; 29,700

D. 27,980;   28,207; 28,097; 29,700

18. Which number correctly completes the subtraction sentence
    5.0 – 3.25 = _____ ?

    A. 1.25
    B. 1.75
    C. 2.25
    D. 2.75

19. Moe packs 55 boxes with flashcards. Each box holds 100 flashcards. How many flashcards Moe can pack into these boxes?
    A. 5505
    B. 550
    C. 275
    D. 5,500

20. A stack of 7 pennies has a height of 1 centimeter. Nick has a stack of pennies with a height of 4 centimeters. Which equation can be used to find the number of pennies, n, in Nick's stack of pennies?

   A. n = 7 + 4

   B. n = 7 − 4

   C. n = 7 × 4

   D. n = 7 ÷ 4

21. What is the value of A in the equation 64 ÷ A = 8

   A. 2
   B. 4
   C. 6
   D. 8

22. Jason's favorite sports team has won 0.62 of its games this season. How can Jason express this decimal as a fraction?

   A. $\frac{6}{2}$

   B. $\frac{62}{10}$

   C. $\frac{62}{100}$

   D. $\frac{6.2}{10}$

23. Use the models below to answer the question.

Which statement about the models is true?

   A. Each shows the same fraction because they are the same size.

   B. Each shows a different fraction because they are different shapes.

   C. Each shows the same fraction because they both have 3 sections shaded.

   D. Each shows a different fraction because they both have 3 shaded sections but a different number of total sections.

24. Sophia flew 2,448 miles from Los Angeles to New York City. What is the number of miles Sophia flew rounded to the nearest thousand?

   A. 2,000
   B. 2,400
   C. 2,500
   D. 3,000

25. Write $\frac{124}{1000}$ as a decimal number.

   A. 1.24

   B. 0.124

   C. 12.4

   D. 0.0124

26. A number sentence such as 31 + Z = 98 can be called an equation. If this equation is true, then which of the following equations is not true?

   A. 98 – 31 = Z

   B. 98 – Z = 31

   C. Z – 31 = 98

   D. Z + 31 = 98

27. Circle a reasonable measurement for the angle:

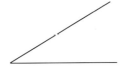

    A. 35°

    B. 90°

    C. 180°

    D. 240°

28. Ella described a number using these clues:
Three-digit odd numbers that have a 6 in the hundreds place and a 3 in the tens place

Which number could fit Ella's description?

    A. 627

    B. 637

    C. 632

    D. 636

29. Tam has 390 cards. He wants to put them in boxes of 30 cards. How many boxes does he need?

   A. 7

   B. 9

   C. 11

   D. 13

30. If this clock shows a time in the morning, what time was it 6 hours and 30 minutes ago?

   A. 07:45 AM

   B. 05:45 AM

   C. 07:45 PM

   D. 05:45 PM

End of the Test

SBAC Practice Test 2

Smarter Balanced Assessment Consortium

Grade 4

Mathematics

2018

- 30 questions
- There is no time limit for this practice test.
- Calculators are NOT permitted for this practice test

1. Which fraction has the least value?
    A. $\frac{1}{2}$
    B. $\frac{3}{8}$
    C. $\frac{3}{4}$
    D. $\frac{9}{16}$

2. Lisa has 336 pastilles. She wants to put them in boxes of 12 pastilles. How many boxes does he need?

    A. 20
    B. 22
    C. 24
    D. 28

3. What is the volume of the cube?

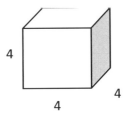

Write your answer in the box below.

[   ]

4. To what number is the arrow pointing?

- A. 26
- B. 28
- C. 30
- D. 33

5. What mixed number is shown by the shaded rectangles?

- A. $3\frac{1}{2}$
- B. $4\frac{1}{2}$
- C. 3
- D. 4

6. There were 12 students in the first row and 8 students in the second row. How many students were in the first two rows?

   A. 20

   B. 24

   C. 88

   D. 96

7. What fraction of each shape is shaded?

   a)
   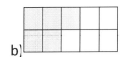
   b)

   A. a. $\frac{5}{16}$; b. $\frac{5}{10}$

   B. a. $\frac{6}{16}$; b. $\frac{5}{16}$

   C. a. $\frac{6}{16}$; b. $\frac{5}{10}$

   D. a. $\frac{8}{16}$; b. $\frac{4}{10}$

8. Joe put 12 red cards and 6 black cards in each bag. What is the total number of cards Joe put in 8 bags?

   Write your answer in the box below.

   ☐

9. What is the perimeter of this rectangle?

   A. 12 cm
   B. 24 cm
   C. 32 cm
   D. 64 cm

10. What is the fourth number in the following pattern?

    1,350, 1,470, 1,590, 1,710, ____, ____, ____, ____

    A. 1830
    B. 1950
    C. 2070
    D. 2190

11. Jamie has 6 quarters, 9 dime, and 11 pennies. How much money does Jamie have?

    A. 150 pennies
    B. 240 pennies
    C. 251 pennies
    D. 281 pennies

12. The temperature on Sunday at 12:00 PM was 76°F. Low temperature on the same day was 24°F cooler. Which temperature is closest to the low temperature on that day?

    A. 76°F
    B. 52°F
    C. 51°F
    D. 75°

13. The number 47.06 can be expressed as _____

    A. $(4 \times 10) + (7 \times 1) + (6 \times 0.01)$
    B. $(4 \times 10) + (7 \times 1) + (6 \times 0.1)$
    C. $(4 \times 1) + (7 \times 1) + (0 \times 1) + (6 \times 1)$
    D. $(4 \times 10) + (7 \times 1) + (0 \times 10) + (6 \times 100)$

14. There are 365 days in a year, and 24 hours in a day. How many hours are in a year?

    A. 2190
    B. 7440
    C. 7679
    D. 8760

15. Jeb paid $72 for a magazine subscription. If he is paying $4 for each issue of the magazine, how many issues of the magazine will he receive?

    A. 18
    B. 20
    C. 22
    D. 24

16. What is the perimeter of the triangle?

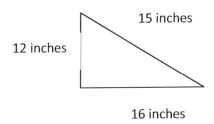

    A. 27 inches
    B. 31 inches
    C. 43 inches
    D. 192 inches

17. Which triangle has one obtuse angle?

A.

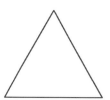

B.

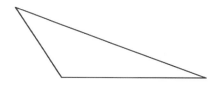

C.

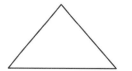

D.

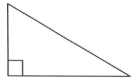

18. A building is 36 feet high. What is the height of the building in yards?

   A. 1 yard

   B. 3 yards

   C. 12 yards

   D. 108 yards

19. The sum of A and B equals 35. If A = 16, which equation can be used to find the value of B?

   A. B − 16 = 35

   B. B + 16 = 35

   C. A + 16 = 35

   D. A − 16 = 35

20. Which number is represented by A?

   9 × A = 108

   A. 9

   B. 10

   C. 11

   D. 12

21. A straight line measures 180°. A straight line and a triangle are touching as shown in the figure below.

What is the value of A in the figure?

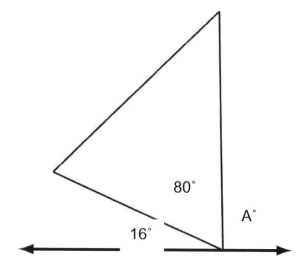

A. 64
B. 84
C. 90
D. 96

22. What is the perimeter of this shape?

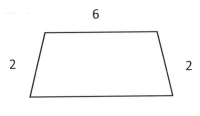

A. 14
B. 48
C. 18
D. 16

23. On Saturday Lily was a referee at 3 soccer games. She arrived at the soccer field 15 minutes before the first game. Each game lasted for $1\frac{1}{2}$ hours. There were 5 minutes between each game. Lily left 10 minutes after the last game. How long, in minutes, was Lily at the soccer field?

A. 300 minutes
B. 305 minutes
C. 480 minutes
D. 485 minutes

24. The figure below shows a diagram of a reading room.

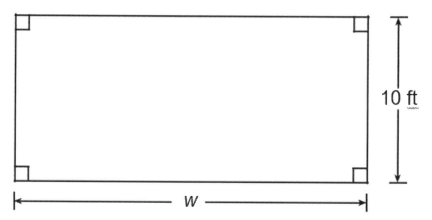

The perimeter of the reading room is 60 feet (ft). What is the width, w, of the reading room?

A. 6 ft
B. 12 ft
C. 20 ft
D. 50 ft

25. Which shape shows a line of symmetry?

A.

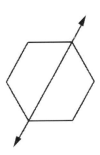

B.

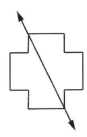

C.

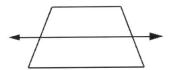

D.

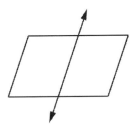

26. Emma draws a shape on her paper. The shape has four sides. It has only one pair of parallel sides. What shape does Emma draw?

   A. parallelogram
   B. rectangle
   C. square
   D. trapezoid

27. Jack has $56.00. He earns $19.00 more. How much money does Jack have in all?

   A. $37.00
   B. $43.00
   C. $65.00
   D. $75.00

28. Rounded to the nearest 10,000, the population of Louisiana was 4,530,000 in 2010. Which number could be the actual population of Louisiana in 2010?

    A. 4,500,321
    B. 4,524,491
    C. 4,533,372
    D. 4,535,343

29. A football teams play 92 games each year. How many games will the team play in 15 years?

    A. 1080
    B. 1140
    C. 1380
    D. 1680

30. Which statement about the number 945,382.16 is true?

    A. The digit 6 has a value of (6×100)
    B. The digit 4 has a value of (4×100)
    C. The digit 8 has a value of (8×10)
    D. The digit 5 has a value of (5×100)

End of the Test

SBAC Practice Tests Answer Keys

SBAC Practice Test 1 Answer Key				SBAC Practice Test 2 Answer Key			
1.	B	2.	C	1.	B	2.	D
3.	C	4.	C	3.	64	4.	C
5.	C	6.	860	5.	A	6.	D
7.	A	8.	A	7.	C	8.	144
9.	A	10.	C	9.	B	10.	D
11.	A	12.	B	11.	C	12.	B
13.	D	14.	A	13.	A	14.	D
15.	C	16.	D	15.	A	16.	C
17.	C	18.	B	17.	B	18.	C
19.	D	20.	C	19.	B	20.	D
21.	D	22.	C	21.	B	22.	C
23.	D	24.	A	23.	B	24.	C
25.	B	26.	C	25.	A	26.	D
27.	A	28.	B	27.	D	28.	C
29.	D	30.	C	29.	C	30.	C

Answers of the worksheets

Place Values

1) 30 + 5
2) 60 + 7
3) 40 + 2
4) 80 + 9
5) 90 + 1
6) ones place
7) tens place
8) ones place
9) tens place
10) hundreds place

Comparing and Ordering Numbers

7) 23 less than 34
8) 89 less than 98
9) 45 greater than 25
10) 34 greater than 32
11) 91 equal to 91
12) 57 greater than 55
13) 85 greater than 78
14) 56 greater than 43
15) 34 equal to 34
16) 92 less than 98
17) 38 less than 46
18) 67 greater than 58
19) 88 greater than 69
20) 23 less than 34
21) -19, -15, -4, 1, 20
22) -5, -3, 2, 4, 6
23) -42, -22, 0, 15, 19
24) -91, -55, -13, 0, 26, 67
25) -71, -54, -39, -25, -17, 90
26) 5, 19, 24, 46, 77, 98

Addition

1) 7,845
2) 8,465
3) 9,451
4) 6,486
5) 5,801
6) 8,404
7) 60
8) 1500
9) 1,100
10) 1300
11) 530
12) 1,560
13) 9,355

Subtraction

1) 3,097
2) 1,891
3) 4,609
4) 2,044
5) 5,461
6) 901
7) 1,103
8) 1,329
9) 546
10) 1,050
11) 1,890
12) 320
13) 1,792

Multiplication

1) 585
2) 320
3) 1,080
4) 2,252
5) 1,825
6) 2,225
7) 16,252
8) 24,856
9) 30,498
10) 120
11) 60

Long Division by One Digit

1) 132
2) 70
3) 29
4) 13
5) 34
6) 21
7) 23
8) 60
9) 16
10) 37
11) 14
12) 40
13) 144
14) 128
15) 142
16) 346
17) 204
18) 853

Division with Remainders

1) 118 R4
2) 98 R1
3) 92 R1
4) 42 R4
5) 146 R2
6) 96 R1
7) 53 R1
8) 410 R0
9) 99 R1
10) 131 R5
11) 161 R3
12) 111 R3
13) 1477 R2
14) 439 R3
15) 336 R3
16) 135 R8
17) 753 R5
18) 1685 R3

Rounding and Estimating

1) 1000
2) 3000
3) 360
4) 70
5) 50
6) 330
7) 1200
8) 9.6
9) 7.5
10) 70
11) 80
12) 90
13) 70
14) 80
15) 40
16) 120
17) 30
18) 130
19) 120
20) 80
21) 90
22) 110
23) 120
24) 780

Odd or Even

1) even
2) odd
3) odd
4) even
5) odd
6) odd
7) even
8) odd
9) even
10) even
11) even
12) odd
13) 22
14) 26
15) 24
16) 58
17) 53
18) 73
19) 63
20) 27

Decimal Place Value

1) one
2) hundredths
3) hundredths
4) tenths
5) tenths
6) thousands
7) hundredths
8) tenths
9) hundredths
10) ones
11) 0.01
12) 10
13) 0.2
14) 700
15) 30
16) 5
17) 0.05
18) 2,000
19) 400
20) 0.2

Order and Comparing Decimals

1) <
2) <
3) >
4) <
5) <
6) >
7) <
8) <
9) >
10) >
11) >
12) =
13) <
14) <
15) >
16) =

17) 0.23, 0.36, 0.4, 0.54, 0.87
18) 1.2, 1.80, 1.97, 2.4, 3.65
19) 0.34, 0.67, 1.2, 1.9, 2.3
20) 1.2, 1.34, 1.7, 3.2, 3.55, 4.2

Decimal Addition

1) 13.36
2) 2.8
3) 8.54
4) 5.15
5) 7.80
6) 10.08
7) 26.7
8) 24.98
9) 28.59
10) 22.80
11) 42.15
12) 29.13
13) 31.09
14) 52.65
15) 31.74
16) 49.4
17) 52.35
18) 45.2

Decimal Subtraction

1) 2.66
2) 1.45
3) 2.11
4) 4.12
5) 2.4
6) 1.23
7) 1.54
8) 0.33
9) 3.38
10) 7.32
11) 2.53
12) 1.8
13) 1.94
14) 1.4
15) 1.76
16) 4.15
17) 4.13
18) 1.84

Money Subtraction

1) 991-761-1319
2) 402-139-443
3) 566-640-1076
4) 4.86

Divisibility Rules

8	2 3 4 5 6 7 8 9 10
1) 16	2 3 4 5 6 7 8 9 10
2) 10	2 3 4 5 6 7 8 9 10
3) 15	2 3 4 5 6 7 8 9 10
4) 28	2 3 4 5 6 7 8 9 10
5) 36	2 3 4 5 6 7 8 9 10
6) 18	2 3 4 5 6 7 8 9 10
7) 27	2 3 4 5 6 7 8 9 10
8) 70	2 3 4 5 6 7 8 9 10
9) 57	2 3 4 5 6 7 8 9 10
10) 102	2 3 4 5 6 7 8 9 10
11) 144	2 3 4 5 6 7 8 9 10
12) 75	2 3 4 5 6 7 8 9 10

Fraction

1) $\frac{3}{8}$
2) $\frac{4}{10}$
3) $\frac{6}{20}$
4) $\frac{4}{42}$
5) $\frac{3}{8}$

Add Fractions with Like Denominators

1) 1
2) 1
3) $\frac{9}{8}$
4) $\frac{6}{4}$
5) $\frac{7}{10}$
6) $\frac{5}{7}$
7) $\frac{8}{5}$
8) $\frac{12}{14}$
9) $\frac{16}{18}$
10) $\frac{8}{12}$
11) $\frac{10}{13}$
12) $\frac{20}{25}$
13) 1
14) $\frac{9}{20}$
15) $\frac{12}{17}$
16) $\frac{23}{32}$
17) $\frac{22}{28}$
18) $\frac{12}{20}$
19) $\frac{35}{45}$
20) $\frac{26}{36}$
21) $\frac{31}{30}$

Subtract Fractions with Like Denominators

1) $\frac{2}{5}$
2) $\frac{1}{3}$
3) $\frac{3}{9}$
4) $\frac{2}{6}$
5) $\frac{1}{10}$
6) $\frac{2}{7}$
7) $\frac{2}{8}$
8) $\frac{2}{13}$
9) $\frac{3}{10}$
10) $\frac{1}{12}$
11) $\frac{6}{21}$
12) $\frac{6}{19}$

13) $\frac{3}{25}$ 16) $\frac{12}{30}$ 19) $\frac{20}{40}$

14) $\frac{18}{32}$ 17) $\frac{5}{33}$ 20) $\frac{20}{35}$

15) $\frac{13}{27}$ 18) $\frac{10}{28}$ 21) $\frac{10}{36}$

Add and Subtract Fractions with Like Denominators

1) 1   8) $\frac{2}{12}$ 14) $\frac{3}{14}$

2) $\frac{5}{6}$ 9) $\frac{21}{25}$ 15) $\frac{3}{15}$

3) $\frac{7}{8}$ 10) $\frac{2}{5}$ 16) $\frac{2}{16}$

4) $\frac{8}{9}$ 11) $\frac{2}{7}$ 17) $\frac{5}{50}$

5) $\frac{5}{10}$ 12) $\frac{1}{4}$ 18) $\frac{3}{21}$

6) $\frac{5}{7}$ 13) $\frac{5}{9}$

7) 1

Compare Sums and Differences of Fractions with Like Denominators

1) $\frac{3}{4} > \frac{1}{4}$ 5) $\frac{2}{9} < \frac{7}{9}$ 9) $\frac{17}{18} = \frac{17}{18}$

2) $1 > \frac{4}{5}$ 6) $\frac{5}{12} > \frac{3}{12}$ 10) $1 > \frac{18}{21}$

3) $\frac{2}{7} < \frac{6}{7}$ 7) $\frac{4}{8} > \frac{1}{8}$ 11) $\frac{10}{16} < \frac{12}{16}$

4) $\frac{16}{10} > \frac{5}{10}$ 8) $\frac{14}{15} > \frac{9}{15}$ 12) $\frac{16}{32} < \frac{20}{32}$

13) $1 > \dfrac{15}{30}$

14) $\dfrac{22}{27} > \dfrac{9}{27}$

15) $\dfrac{27}{45} < \dfrac{30}{45}$

16) $\dfrac{47}{36} > \dfrac{18}{36}$

Add 3 or More Fractions with Like Denominators

1) 1

2) 1

3) $\dfrac{7}{9}$

4) $\dfrac{3}{4}$

5) $\dfrac{14}{15}$

6) $\dfrac{8}{12}$

7) $\dfrac{7}{10}$

8) $\dfrac{13}{18}$

9) $\dfrac{19}{21}$

10) $\dfrac{15}{16}$

11) $\dfrac{12}{25}$

12) $\dfrac{24}{30}$

13) $\dfrac{21}{27}$

14) $\dfrac{14}{42}$

Simplifying Fractions

1) $\dfrac{11}{12}$

2) $\dfrac{4}{5}$

3) $\dfrac{2}{3}$

4) $\dfrac{3}{4}$

5) $\dfrac{1}{3}$

6) $\dfrac{1}{4}$

7) $\dfrac{4}{9}$

8) $\dfrac{1}{3}$

9) $\dfrac{5}{2}$

10) $\dfrac{3}{27}$

11) $\dfrac{5}{9}$

12) $\dfrac{3}{4}$

13) $\dfrac{5}{8}$

14) $\dfrac{13}{16}$

15) $\dfrac{1}{5}$

16) $\dfrac{4}{7}$

17) $\dfrac{1}{2}$

18) $\dfrac{5}{12}$

19) $\dfrac{3}{8}$

20) $\dfrac{1}{4}$

21) $\dfrac{5}{9}$

Add fractions with unlike denominators

2) $2\frac{2}{3}$

3) $\frac{14}{15}$

4) $\frac{4}{3}$

5) $2\frac{11}{36}$

6) $\frac{3}{5}$

7) $\frac{13}{14}$

8) $1\frac{3}{20}$

9) $\frac{13}{15}$

10) $\frac{33}{25}$

11) $\frac{7}{6}$

12) $\frac{3}{4}$

13) $\frac{26}{15}$

Subtract fractions with unlike denominators

1) $-\frac{6}{5}$

2) $-\frac{13}{14}$

3) 0

4) $\frac{13}{45}$

5) $\frac{3}{14}$

6) $\frac{1}{6}$

7) $\frac{1}{36}$

8) $\frac{9}{40}$

9) $\frac{7}{18}$

10) $\frac{6}{25}$

11) $\frac{8}{27}$

12) $\frac{5}{21}$

13) $\frac{1}{3}$

14) $\frac{7}{20}$

Add fractions with denominators of 10 and 100

1) $\frac{7}{10}$

2) $\frac{11}{20}$

3) $\frac{17}{20}$

4) $\frac{83}{100}$

5) $\frac{22}{25}$

6) $\frac{4}{5}$

7) $\frac{9}{10}$

8) $\frac{4}{5}$

9) $\frac{89}{100}$

10) $\frac{41}{50}$

11) 1

12) $\frac{7}{10}$

13) $\frac{19}{25}$

14) $\frac{87}{100}$ 16) $\frac{19}{20}$ 18) $\frac{24}{25}$

15) $\frac{17}{20}$ 17) $\frac{77}{100}$

Add and subtract fractions with denominators of 10, 100, and 1000

1) $\frac{50}{100}$ 7) $\frac{6}{25}$ 13) $\frac{2}{5}$

2) $\frac{87}{100}$ 8) $\frac{9}{50}$ 14) $\frac{4}{5}$

3) $\frac{7}{10}$ 9) $\frac{33}{25}$ 15) $\frac{21}{50}$

4) $\frac{41}{100}$ 10) $\frac{91}{100}$ 16) $\frac{9}{50}$

5) $\frac{23}{25}$ 11) $\frac{29}{50}$ 17) $\frac{1}{5}$

6) $\frac{59}{50}$ 12) $\frac{1}{4}$ 18) $-\frac{4}{25}$

Fractions to Mixed Numbers

1) $2\frac{1}{4}$ 7) $1\frac{2}{3}$ 13) $3\frac{1}{2}$

2) $7\frac{2}{5}$ 8) $1\frac{4}{5}$ 14) $9\frac{3}{4}$

3) $3\frac{1}{2}$ 9) $3\frac{4}{5}$ 15) $7\frac{1}{5}$

4) $4\frac{1}{10}$ 10) $2\frac{7}{10}$ 16) $4\frac{1}{3}$

5) $5\frac{1}{2}$ 11) $1\frac{2}{3}$ 17) $5\frac{5}{8}$

6) $5\frac{3}{5}$ 12) $2\frac{1}{8}$ 18) $5\frac{2}{5}$

4th Grade SBAC Math Workbook 2018

Mixed Numbers to Fractions

1) $\dfrac{4}{3}$

2) $\dfrac{8}{3}$

3) $\dfrac{16}{3}$

4) $\dfrac{34}{5}$

5) $\dfrac{11}{4}$

6) $\dfrac{19}{7}$

7) $\dfrac{32}{9}$

8) $\dfrac{29}{10}$

9) $\dfrac{47}{6}$

10) $\dfrac{83}{12}$

11) $\dfrac{169}{20}$

12) $\dfrac{42}{5}$

13) $\dfrac{29}{5}$

14) $\dfrac{55}{6}$

15) $\dfrac{15}{4}$

16) $\dfrac{32}{3}$

17) $\dfrac{51}{4}$

18) $\dfrac{104}{7}$

Add and Subtract Mixed Numbers with Like Denominators

1) $13\dfrac{18}{20}$

2) $10\dfrac{13}{10}$

3) $13\dfrac{3}{4}$

4) $13\dfrac{5}{6}$

5) $3\dfrac{2}{3}$

6) $5\dfrac{4}{5}$

7) $4\dfrac{7}{9}$

8) $1\dfrac{9}{10}$

9) $3\dfrac{24}{25}$

10) $10\dfrac{7}{6}$

11) $5\dfrac{4}{8}$

12) $8\dfrac{6}{10}$

13) $8\dfrac{6}{12}$

14) $5\dfrac{6}{9}$

Order of Operations

1) 9

2) 15

3) 32

4) 17

5) 33

6) 56

7) 33	11) 29	15) 28
8) 49	12) 9.8	16) 10
9) 6	13) 12	17) 50
10) 2	14) 62	18) 18

Line Segments

1) Line segment
2) Ray
3) Line
4) Line segment
5) Ray
6) Line
7) Line
8) Line segment

Identify lines of symmetry

1) yes
2) no
3) no
4) yes
5) yes
6) yes
7) no
8) yes

Count lines of symmetry

1)
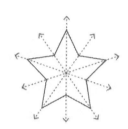

2)
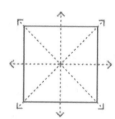

3)
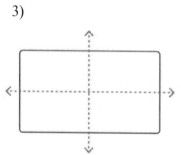

4)
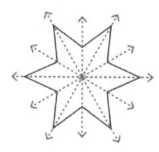

5)
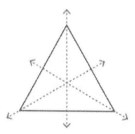

6)
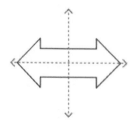

7)
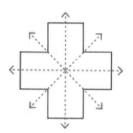

8)
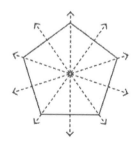

Parallel, Perpendicular and Intersecting Lines

1) Parallel
2) Intersection
3) Perpendicular
4) Parallel
5) Intersection
6) Perpendicular
7) Parallel
8) Parallel

Identifying Angles: Acute, Right, Obtuse, and Straight Angles

1) Obtuse
2) Acute
3) Right
4) Acute
5) Straight
6) Obtuse
7) Obtuse
8) Acute

Angles of 90, 180, 270, and 360 Degrees

1) 90°
2) 180°
3) 270°
4) 360°

Measure Angles with a Protractor

1) 90°
2) 30°
3) 120°
4) 150°

Polygon Names

1) Triangle
2) Quadrilateral
3) Pentagon
4) Hexagon
5) Heptagon
6) Octagon

Classify triangles

1) Scalene, obtuse
2) Isosceles, right
3) Scalene, right
4) Equilateral, acute
5) Scalene, acute
6) Scalene, acute

Parallel Sides in Quadrilaterals

1) Square
2) Rectangle
3) Parallelogram
4) Rhombus
5) Trapezoid
6) Kike

Identify Parallelograms

1) Rhombus
2) Squares
3) Rectangles
4) Parallelogram

Identify Trapezoids

Number of 1, 4, 5, 7

Classify Quadrilaterals

1) Square
2) Rectangle
3) Trapezoid
4) Parallelogram
5) Trapezoid
6) Kite

Identify Three-Dimensional Figures

1) Cube
2) Triangular pyramid
3) Triangular prism
4) Square pyramid
5) Rectangular prism
6) Pentagonal prism
7) Hexagonal prism

Count Vertices, Edges, and Faces

		Number of edges	Number of faces	Number of vertices
7)		6	4	4
8)		8	5	5
9)		12	6	8
10)		12	6	8
11)		15	7	10
12)		18	8	12

Identify Faces of Three-Dimensional Figures

1) 6
2) 2
3) 5
4) 4
5) 6
6) 7
7) 8
8) 5

Telling Time

1) 12:00
2) 22:10 pm
3) 3 hours and 20 minutes
4) 2 hours and 35 minutes
5) 3:30 pm

Tally and Pictographs

fish	☺☺☺
frog	☺ ☺ ☺
sheep	☺☺☺
butterfly	☺
squirrel	☺☺☺☺

Bar Graphs

Line Graphs

1) 3
2) Thursday
3) Wednesday
4) 26

Patterns: Numbers

1) 3, 6, 9 , 12, 15, 18, 21, 24
2) 2, 4, 6, 8, 10, 12, 14, 16
3) 5, 10, 15, 20, 25, 30, 35, 40
4) 15, 25, 35, 45, 55, 65, 75, 85
5) 11, 22, 33, 44, 55, 66, 77, 88
6) 10, 18, 26, 34, 42, 50, 58, 66
7) 61, 55, 49, 43, 37, 31, 25, 19
8) 45, 56, 67, 78, 89, 100, 111, 122

Perimeter: Find the Missing Side Lengths

1) 11
2) 9
3) 5
4) 4
5) 15
6) 4
7) 7
8) 9

Perimeter and Area of Squares

1) A: 25, P: 20
2) A: 9, P: 12
3) A: 49, P: 28
4) A: 4, P: 4
5) A: 16, P: 16
6) A: 100, P: 40
7) A: 64, P: 32
8) A: 144, P: 48

Perimeter and Area of rectangles

1) A: 50, P: 30
2) A: 24, P: 20
3) A: 21, P: 20
4) A: 150, P: 55
5) A: 55, P: 32
6) A: 72, P: 34
7) A: 32, P: 24
8) A: 72, P: 36

Find the Area or Missing Side Length of a Rectangle

1) 70
2) 6
3) 10
4) 96
5) 330
6) 30
7) 12
8) 25

Area of Complex Figures (With All Right Angles)

1) 16 cm^2
2) 360 cm^2
3) 112 cm^2
4) 6144 cm^2

Area and Perimeter: Word Problems

1) 9
2) 24
3) 30
4) 256
5) 121
6) 26

Area and Perimeter to Determine Cost

1) 9
2) 24
3) 30
4) 256
5) 121
6) 26

Calculate Radius, Diameter, and Circumference

1) diameter: 14 circumference: 43.96
2) diameter: 16 circumference: 50.24
3) diameter: 24 circumference: 75.40
4) diameter: 32 circumference: 100.53
5) radius: 11
6) radius: 18
7) radius: 23
8) radius: 12

Measurement - Time

1) 3:20;
2) 2:25
3) 1:25
4) 1:15
5) 30;
6) 25
7) 900
8) 660
9) 1,620
10) 600
11) 1,200
12) 720

Measurement - Metric System

1) 4 mm = 0.4 cm
2) 0.6 m = 600 mm
3) 2 m = 200 cm
4) 0.03 km = 30 m
5) 3000 mm = 0.003 km
6) 5 cm 0.05 m
7) 0.03 m = 3 cm
8) 1000 mm = 0.001 km
9) 600 mm = 0.6 m
10) 0.77 km = 770,000 mm
11) 0.08 km = 80 m
12) 0.30 m = 30 cm
13) 400 m = 0.4 km
14) 5000 cm = 0.05 km
15) 40 mm = 4 cm
16) 800 m = 0.8 km

Measurement – Length

1) 24
2) 60
3) 3
4) 9
5) 100
6) 3000
7) 10000
8) 24

Measurement - Volume

1) 1344 cm^3
2) 1650 cm^3
3) 512 m^3
4) 1144 cm^3
5) 36
6) 44

Measurement - Temperature

1) 20
2) 50
3) 95
4) 176
5) 59
6) 77
7) 122
8) 113
9) 194
10) 86
11) 68
12) 158
13) 176

Statistics

1) mean: 3, median: 2, mode: 1.2, range: 6
2) mean: 3.625, median: 3, mode: 2, range: 5
3) mean: 5.22, median: 4, mode: 4, range: 8
4) mean: 4, median: 4, mode: 4, range: 6
5) mean: 7, median: 7, mode: 5, 7, 8, range: 4
6) mean: 4.2, median: 4, mode: 1,2,4,5,9, range: 8
7) mean: 5, median: 5, mode: 5, range: 8
8) mean: 5.78, median: 6, mode: 7, range: 7
9) mean: 5, median: 5, mode: 2, 4, 5, 6, range: 7
10) mean: 6.125, median: 5, mode: 5, range: 8
11) mean: 3, median: 2, mode: 2.5, range: 4
12) mean: 5, median: 5, mode: none, range: 7

Roman Numerals

1) II
2) VI
3) IV
4) IX
5) X
6) VII
7) III
8) I
9) V
10) VIII
11) IX
12) XI
13) VI
14) XII

Patterns

6, 3, 12

498, 483, 468

40-48-56-64-72-80

23-34-45-56-67-78

38-45-52

38-31-24

30,23,16

25,12, -1

1950-2150-2350

61

"Effortless Math" Publications

Effortless Math authors' team strives to prepare and publish the best quality Mathematics learning resources to make learning Math easier for all. We hope that our publications help you or your student learn Math in an effective way.

We all in Effortless Math wish you good luck and successful studies!

Effortless Math Authors

Online Math Lessons

Enjoy interactive Math lessons online
with the best Math teachers

Online Math learning that's effective, affordable, flexible, and fun

Learn Math wherever you want; when you want
Ultimate flexibility. You can now learn Math online, enjoy high quality engaging lessons no matter where in the world you are. It's affordable too.

Learn Math with one-on-one classes
We provide one-on-one Math tutoring online. We believe that one-to-one tutoring is the most effective way to learn Math.

Qualified Math tutors
Working with the best Math tutors in the world is the key to success! Our tutors give you the support and motivation you need to succeed with a personal touch.

Online Math Lessons

It's easy! Here's how it works.

1- Request a FREE introductory session.

2- Meet a Math tutor online.

3- Start Learning Math in Minutes.

Send Email to: info@EffortlessMath.com

Or Call: **+1-469-230-3605**

Made in the USA
San Bernardino, CA
03 June 2018